LESSONS IN CAMOUFLAGE

Poems

Martin Ott

C&R Press
Conscious & Responsible

First Edition
1 2 3 4 5 6 7 8 9

Selections of up to two pages may be reproduced without permissions. To reproduce more than two pages of any one portion of this book write to C&R Press publishers John Gosslee and Andrew Sullivan.

Cover Art "The Conquest of Nature" by Eugenia Loli
Interior Design by Ali Chica

Library of Congress Cataloging-in-Publication Data

ISBN 978-1-936196-67-8
Library of Congress Control Number: 2018932522

C&R Press
Conscious & Responsible
crpress.org

For special discounted bulk purchases please contact:
C&R Press sales@crpress.org

"Three things cannot be long hidden: the sun, the moon, and the truth."

-Buddha

LESSONS IN CAMOUFLAGE

CONTENTS

The King of Camouflage

Yesterday's sky is my molting skin.
My soldier's makeup stains the plays
on a global stage, the actors dead, maudlin
applause sweeping across your family's strays.
See the tapestry of faces flickering in
the cracks of the windshield, the haze
of suffering to raise your children
to crush third grade, seize glory days.
God's oversized disguise is my old-school spin
of the head to jumble the most recent phase
of who you're supposed to be, the only sin
giving control to me, the king you chase
with suit and tie, power strut, my glittering ring
on your finger a distraction from the craze
of waking in a cold swear, fear of drowning
the spectacular affixing of my ways.
Look into ash. I'm there where you begin.
I am the shadow that forms in front and stays
long after you've lost the spark within.
My power plumes in endless praise.

Marks

Remember the guy on the infomercial
waving the wand over impossible stains,
erasing kids' mistakes, impromptu
graffiti flayed into paint, the ottoman
saved, time itself swiped away.
Remember how we laughed stupidly
on the front porch couches in winter
skipping classes, waiting for the cars
to spin out, the crash of vehicles,
the grace to care spooling in smoke.
Remember the scars of night sports,
the tattoo of a sanguine penguin
brandishing a machine pistol,
the knot of fury beneath shoulder
blades, territorial males in bloom.
Remember the planet that disappeared
from textbooks, the moon mantra
that boiled away in car exhaust,
years of maneuvering among a cadre
in khakis and collars, the rub of self.
Remember what it is to remember
the lakes of childhood, the hands
groping beneath the waves. Serpents
of the deep are rumored to peer up
with fish eyes, our own refracted view.
Remember when the weight of children
on your shoulder indented your life,
the fears of childhood and loss
of children now linked in the moss
of your daydream, the uprooted tree.
Remember the age spots on your hands
unidentifiable from your lost mother.
Remember the night she stopped
breathing, the string that connected
you to earth snipped, centrifuge of grief.
Remember the sting from the first woman
who slapped you besides your mother,

the cascade of hope and terror beyond
your ability to control, birth of manhood
itself the more painful passage.
Remember that waking is an act
of defiance, the creases of history
flipped open in the here and now,
the conundrum of who and how,
attainable in the kingdom of why.

Mile Post

I made sure that no one ever passed me
in the two mile race on desert paths.
Why am I still running from camouflage?
The Army was the first time I knew.
In the two mile race on desert paths,
the way a space shuttle slaloms before coming apart,
the Army was the first time I knew
other soldiers sprinting from the partners they'd lost.
The way a space shuttle slaloms before breaking apart,
I raced from a bully chasing me from the school bus,
other soldiers sprinting from the partners they'd lost.
I fought through the ring of children looking for blood.
I raced from a bully chasing me from the school bus;
the sling of the rifle I fired still weighs on me.
I fought through the ring of children looking for blood.
Fear closes ground only when you turn to fight it.
The sling of the rifle I fired still weighs on me.
The Boston Marathon blew open the wall of the world.
Fear closes ground only when you turn to fight it.
Night will never catch day on the cratered runway.
The Boston Marathon blew open the wall of the world.
I made sure that no one ever passed me.
Night will never catch day on the cratered runway.
Why am I still running from camouflage?

Invisible Monsters

The worry was still there, but I was gone.
I fled gasping for air, but I had no one
to hold during the monsoon, malevolence of myself.
Let's turn back the clock. Easy enough. Filet
of man, panic attacks that shook the foundation
behind each door, the specter of falling apart never
far away. *The Mastery of Oneself* is not a tome
for the poet's bookshelf or the tycoon's bedside,
belief in the imminence of pen and word
no match for the snail, the crunch beneath
boots, the trigger of each day, phalanx
of friends no shield for a planet made thin
by those who aim to swallow it like a lozenge.
My son has a nightlight and I remember how
the invisible monsters kept me in their thrall,
the attempt to burn away shadows from sin.
I should have noticed the man-shaped fear
following me, the odd resemblance of bosses
to Bin Laden, the bogie man of the moment
gripping a father's belt and the crackle of life
slipping away with tonic and gin. Soldiers are sold
and the rich men grow bold in the distant haze
of another spin. I hold my son close sometimes
so he can sleep and reach out my hand for the moon
soaked in sun, a way to delay the beating of wings.

Uniform

The white flag of surrender
blended in, the bleached
fatigues, the skeleton key
beneath the threads.
The office is now the battlefield.
Our hot date is an interview.
The barista is a fund manager.
Our mistake was the shedding.
The inevitable routine wore us,
an illusion of sameness.
The fabric flutters on a pole
that used to be a man.

Why My Father Carries Three Guns

Who will be my brother in a family of guns?
The virtuoso performance is the one not given in a symphony of guns.
Nothing creates more holes than earthworms.
We become rigor mortis and sandstone, skin casing, necrobiology of guns.
If a gun goes off in the forest and no one hears?
Gandalf teeters on a geoglyph with a thunderstick, the shamanism of guns.
Paralysis is permanent and vestments are hidden.
In holy places, the myth of flooding is the flooding of guns.
String theory places the bullet in a state of perpetual being.
Our schools teach us the tragic trigonometry of guns.

The Gravel Diaries

The pen scratches a long-ago itch.
A one-eared dog brays at a coyote
invading his street. The delivery
truck coughs too close for comfort.
I hid away in my room, lost in yellow,
the light stabbing villains and time-
washed pages. A child's toy dagger
hisses in the scabbard. The LA River
gurgles in a tectonic bouillabaisse.
My heroes for a time hid in spines.
Damp shirts on the balcony ululate
on a swinging noose. An inconsolable
lover sighs only within earshot.
The wail follows me from when my husky
was put to sleep. The smart phone quakes
in the middle of the night. Shrapnel
whistles for its fleeing companions.
I read one book for each time I cleaned
my rifle. Gravel grinds its endless migration
back home. Friends were lost to distance,
to madness, to drugs and to the ditch
I tossed things into when I fled the scene.
Losses pile up, rumble from yet another
subterranean port. Passage is paramount.
Books saved me from the abyss.

Prayer For Morning Commute

Look out for the Janus behind the wheel,
the multitasker sexting or crimping lashes,
the eater smoker and drinker toker, the mad
flipper and the distracted talker in all forms.
I once thought God made me a wind talker.
Give me the patience to not pray for painful
karma to catch up to legions of near crosswalk
crossers whose mission to save a few seconds
places everyone in danger of near catastrophe.
The ultimate traveler swirling everywhere.
Help me feel compassion for the Filipinotown
man wandering the streets between gas station
and salsa bar hauling a wooden cross to work,
in fear of it being stolen or exercising demons.
Pages flapping in that first angry testament.
Make me gasp in wonder at the several dozen
drivers apparently teleported from some alien
world and placed for the first time in drivers'
seats to weave between interdimensional lanes.
Lessons of the kind wanderer unanswered.
Find someone to pull the wires from my horn
that trumpets its glory far too often for mortal
souls as the lesson of driving cannot be passed
on with a single note sustained for all eternity.
The yearning I had to connect with him.
Save the bird from its unerring path to roadside
meltdown, the singular finger of the motorways
too easily cast and resulting punching of pedals
not for the weak of heart or the meddling spirit.
The journeys from the couch brought me farther.
Most of all, please assist those who have stopped
on their path to righteousness to decide between
the crimson fires of the brake lights and angelic
spots beaming from those who've found a way.
The search for a father carried me further still.

Camels Of The American West

Some are rumored to be descended
from the first camels that spawned
Sahara cousins, passage on the land
bridge back to the old world, spit
into the handshake to consummate
brothers the camel's way of phlegming
friend and foe. Vessels hauled camels
along with other slaves, winds shifting
armies from dunes to merchant monsoons.
Dun deserters from the Camel Corps fled
across the Rio Grande devouring cactus
blooms. The legend of the Red Ghost
haunted the plains, blamed for a trampled
woman's death, a camel with a young
soldier, now skeleton, tied to its back
as a riding lesson. The evidence: red fur,
jangling bones, real fear. The first movie
stars rode camels along the dirt roads
of Los Angeles, the oasis a parasite.
We look to the deserts of America
for aliens of all kinds, rumors of UFOs
and camel herds resistance to extinction,
blending into the rain of experimental
rockets scorching sand, time released
in curses and unspun campfire verses.

I Lost The Robot In The Divorce

With one eye, tin torso, dryer duct limbs,
my thrift store doppelganger leans over
the scratched-up dining room table, dressed for
holidays I cannot view except dim
dreams through cyborg senses. Dressed in elf hat
for Christmas, next to my empty station
guests comment about the combination
lock on the broad metal chest welded shut.
Have I been replaced by this golem pop
clothed in the costumes my kids and I wore
on the days before the shades held races?
Looking back, there was no way I could stop
things splitting, division of flesh, curse
of an immobile man in two places.

The Island Of Escalators

Is the land where no one looks
down. There are motors within
motors, an engraving of jumping
beans on the bottom or so legend
rolls. Who can remember the days
before endless lift, serrated steps,
the apostles of past eras fueling
an epoch? Those who no longer
fear the crack monster catalog
their blasphemies as punishment,
dreams of falling cobbled in air,
songs passed upward, hymn spiral.
This freedom is centipedal, progress
for the sake of it, this balancing
act of heaven, the fear of making it.

Shelter

Never was there a time when
the pantry doors did not bang
from sleepwalkers, human winds.
My earliest memory of home
was the urge for departure,
the creaking of a house easing
into middle age, subtle disarray,
damning evidence of mortality.
I traveled the country searching
for the center, the idea of home.
The myth of the cave captured
stick figures, fattened with time
on rectangles hanging on wire.
I began to burrow into a place
only after the birth of my kids.
We tell tales of those who flew
into mire, the sun, San Dimas:
those without bread float away.
The landscape carries memories,
sure, but lovers hold home within.
We clip our nails to convince
ourselves we no longer rend
and spear shadowy remains.
The last lesson of being home
is to let go of the need of place.
The box around us whispers
in the quiet hours of distance
within us, walls that hold us.

To The Guy Who Drew A Penis On The Elevator

Thanks for giving us something to look
at when my kids visit, for the devotion
it took to bring a chair to etch it
near the ceiling, or else your ability
to dunk without jumping, a giant among
us, armed with pen. The smiley face
on the head was a nice touch. Kudos
to sharing your big ideas with joy.
Could you also be the one who rips
the signage to the courtyard down
each day with Herculean effort,
or pops out the steel ceiling panels?
Sorry to assume that you're a man,
but the shadows in my brain depict
a silhouette with heart-shaped balls.
Are you teaching us no container is
permanent, from womb to coffin,
that the journey homeward is a messy
business? When the floor is damp,
I wonder if your eyes leak the myth
of creation even as you inspire us
to hold our children close in this
shivering box that squeaks for us
to pay attention to the passage.

Hide And Seek

There was that epic game
where you lost yourself.
It started when you slipped
between shower curtains
and we called the police.
Not even tracking dogs
could follow your descent.
The rumors became our oxygen.
You plucked the banjo on Fuji,
dodged venomous snakes in Perth.
You delivered babies in Iceland
Ubers and shrank to toy soldier
size to sabotage drone strikes.
You were missing in action
long before we started looking.
We'd lost you over the years,
one hiccup, one binge, one
scar too many above the eye.
The trigger reflex was ours
under covers and our couches
could not hold your weight.
You could only swipe at
ghosts so long until we were them.
You were just following the rules
that you needed to be found.
You follow us in beltway whispers,
shadows lodged in cushions,
wars without beginning or end.

Why I Worry My Mom Is Dying,
Explained By Five Extinct Punctuation Marks

Manicule

With finger pointer, email flail, I blamed
my mom for passing her guilt onto me,
my daughter. Now indexed in the margins.

Percontation Mark

Are you dying is not rhetorical.
Jagged letters cut. Question are not
scythes. Tilled lines. Are you there?

Pilcrow

The cigarette hack. Rattling of lungs.
The highlighter, human paragraphos.
The X-ray with spots, imagined doom.

Interrobang

The subsequent voicemails. Minute rants.
Affirmation of care, and dispersed fury.
The beep after my harangues. Unaware.

Virgule

The pause while I wait to hear
could be an instant, or forever,
held breath, smoke, fire unclear.

Stranger

The man emerged from a Starbucks
on a mission with moon-landing steps
and a boom box blasting Foreigner,
a table of businessmen foot tapping
to *I Want to Know What Love Is.*
Dude rocked a mustache, orange
vest with no undershirt, skinny
jeans, flip flops, a vision of past
fashion failures or an apocalyptic
future. He cradled a rectangular
slab of gray metal in his other arm
that could have been the fuel
cell to his space ship, offspring
of a departed cyborg companion,
the missing item from every piece
of Ikea furniture, the thing surely
out of place on a man out of time.
His destination was not my place,
but the wake was felt from the top
floor to those in utero. The song
vibrated in the office park, rattled
windows and passersby, Pied Piper
in reverse, the parting of possibilities.
Maybe we're all singing to ourselves,
castaways on a voyage we dare not,
trapped in a past of our construction,
unreachable in the lunchtime bustle.

The Human Library

Who can forget the ending of *Fahrenheit 451*,
campfire tales reworked by Brothers Grimm,
the rhymes to remember, genesis of epochs?
I once got on Wikipedia with John and Andy,
and we wrote outlandish facts about Ayn Rand:
Hitler's lover, school bus driver, Tae Bo instructor.
In nameless crypts all over the world, prisoners
are tagged and warehoused by the hundreds,
referenced by interrogators to recheck the facts.
The collective storyteller unearths our history
from shards of arrows, scrolls, and lost kings,
but who is really behind the unmaking of things?
I am addicted to writing with others, creation
inseparable from friendship, this act a reservoir
that pumps to and from a writer's haunts.
The first books were probably made of skin,
early dissertations of how much trouble
we were in, a lost pantheon inventing gods.
The Odyssey is whispered in the dark at hostels.
We change the narrative when it suits us.
Which reads are bound, and which are boundless?

Calling The Time

In the operating room, the surgeon
hung clocks from all over the world
to the proper time, the tick in other
places to remind her of the worlds
beyond the body, past the expiration
dates. She imagined hand-painted
cuckoos beside *maneki neko* clocks,
the din of metal in fingers clicking
with the promise of good fortune.
She remembered grandfather's fob,
timepieces handcrafted from bone,
plastic mushroom clocks from swap
meets, Elvis clocks and football clocks,
a clock of the moon landing beside
a cow-jumping-over-the-moon clock.
Her mother once told her God's second
hand is the moon, the rhythm of blood
beyond our control. When the heart stops
even light is a clock to the last witness,
the space in between breathing places.
On the door to where the family waits,
she'd set the wristwatch of her patient
on the doorknob to keep them honest
in departure, the alarm of precision,
the telltale of the need to remember.

Imagining The Mouth Of My Sleeping Lover Holds The Power Of Good And Evil

This is the space
between sea and mountain,
green tendrils poking
into heaving clouds,
secret ingredients
extracted from a rare
Amazonian pill
bug and the night sweats
of a dreaming woman.
Those who know
the formula have been
rumored to hold one
finger in the air,
wet with spittle,
to find the winds
that will whisk crumbs
and sand over the latest
dying empire. You need
to cook up a language
no one else knows,
no invented love
monikers or Latin
knock-off, the spell
of a ghost civilization
looking to bind
the world, plucked
promises in a kettle.
So many have entwined
limbs and lambs,
mortared generations
into wheat flour,
boundaries of sheets,
tears held back
to turn one sip
into a lasting testament.
Nothing can make you

live forever in a world
baking with sighs
and dim recollections
except perhaps a kiss
from one to all,
evaporating loss.

Traffic Songs

We left the stove cackling, the door unlocked.
Our children wandered. How they wandered.
Sidewalks under the overpasses swelled.
The land was hungry to reclaims manors.
The tigers remembered how magicians tasted.
Work became a way we inwardly deconstruct.
A roof is a concept. A roof of stars a conceit.
Wheels stretch from wrists. This life. This hunt.
Selfies in museums. Signs to make us think we
are. Rides were our surfboards. Separation
from sharks. Our freedom was the illusion
of mobility, the nostalgia of traffic songs.
Remember how our clocks moaned? Crosses
pointed at Mars? The copter clacking above
above gridlock? We think we must swim
or we will die but the truth is a cresting
wave, or the land collapsing, or crackling
guns, the mistakes of backfire, the hum.

The Mystery Of Three Things

What is with the surfboard, Samsonite
suitcase and Hendrix poster abandoned
in the apartment complex hallway,
no note or explanation, castaways?
The affinity of doorside donations
is the mystery of like minds bridged.
The surfboard has a story to tell,
confidence lost in swells, gray figure
scraping coral, curses like all ocean
fugues, a whisper. What is it about
me that assumes sadness and things
broken that cannot be replaced?
The axiom of the old man and the sea
is the mystery of castles flattened.
The mottled suitcase is the kind
abused by gorillas in commercials,
still holding secrets into old age,
the shell of an adventurous self.
The dereliction of world travelers
is the mystery of intermediate family.
My stepfather left on business trips,
always with one carry-on, children
looking for the souvenirs of staying.
The insurrection of shelved music
is the mystery of tunes coming true.
The Hendrix poster is a close-eyed
portrait surrounded by oval swirls,
from a time that may have never
existed, a lost lover chasing me
across canyon lands to the tune
of *Bad Boys Play Nice.* Hurry.
These things phase and flutter
with no angel woman to rescue me.
The desire to embrace the scattered
is the mystery of holding on.

Mr. Old Year

Parked the murdered-out skeleton
of his automatic car in the front
yard on cinderblocks, the disembodied
computer voice in countdown mode,
the number of times we throttled our love.
In the gated complex, a delivery drone
dropped a package on the doorstep,
bottled water from the last glacier,
models scattering beside a pool,
empty except for your scattered things,
heated with spent and bent titanium rods,
tsunami-irradiated and sizzling backdrop
for selfies, mojito in the foreground,
recreational vehicle rocking behind,
all the voices reminding me of you.
Electronic cigarettes twirled in Gatsby's
fingers, uninvited guests unimpressed
with tiny phalluses, breathless and deathless,
no way to connect the clouds with the sky,
two people on other sides of the parapet.
The partygoers set the host ablaze,
on the final ticks with ash billowing
from the bowl of the desert city,
with a dizziness resembling snow,
a readiness to become the rain.

Morels

When I was a boy, my family and I took
long forays into the woods for berries,
Dachshund in tow, pinging our haul
into pails, sometimes searching for morels.
Mom's body is pale, tumors nestled between
windpipe and heart, five days since she collapsed.
She canned the harvest every summer, thrifty
from growing up poor, father who molested
her. *So many secrets in eyes fluttering open for days,*
even after doctors said her brain no longer worked.
We dug through the underbrush for mushrooms,
among beer cans with bullet holes left by hunters.
I swashbuckled flies and tree arms with branches.
We held her hands and asked doctors to stop
the grimace emanating from the ancient part
of her brain. She only told my sister the story
of why she fled at sixteen with a man she barely
knew, the sound of belts and her mother's guitar
on the porch fading into the light. *We sat with her*
for two days until her heart followed thoughts into
a place, a time, a new home. I always sensed
something lurking beneath the ferns as summers
coiled and I drifted, bucket overturned, to hunt
for meaning in shade, the wrinkles of the earth.

Pilates Instructor Calls 911 After Mistaking Labradoodle For Lion

Perhaps the doors to the fitness
center should not have been left
ajar to let in air, and the dog's owner
should have better tied the rhinestone
leash to a lamp pole. The tan mane
was freakishly puffy, and reminiscent
of a disfigured lion king. It was difficult
to make out details in the chaos of yoga
pants and shrieks. The patrons were
frightened by endless tales of rampage,
and the instructor led his class into
the closet, hero that he was, assured no
feline possessed opposable thumbs
to turn the knob. They tried not to fret
about the lack of oxygen in the darkness
or the receptionist getting maimed,
resisting impulses, sharing bareness,
lit by smart phones, bodies pressed,
an excited beast chasing a giant ball
over matts made on distant shores.

Last Days Of Peekaboo

My vessel on this voyage
has scabs healed to scars,
skin hardened to gastropod,
blemishes there as scripture.
Fear: my heart is a prism.
My kids are portmanteaus
from writers with divergent
stories, their home a pestle
filling with a dash of this, that.
Question: how to gauge childhood?
Warnings spill out sometimes,
words overboard, bobbing buoys
to mark exposed reefs, sharp teeth.
Life is not a hook to be baited.
Desire: to let mistakes be. Human.

Policy

There is a sky that is not a sky,
a roof, perhaps, tumbling down,
the firmament torn, prayers as policy.
There's a famous cricketer who insured
his bushy mustache against incendiary fans,
depilatory kisses, and Edward Scissorhands.
There are distant towns that offer protection
against mangos raining through windshields,
ghost infestations and immaculate conception.
There are meteoric premiums for body parts:
actors' asses, quarterbacks' canons, a yo-yo
master's mitts, a tart food critic's taste buds.
There is someone, somewhere, looking to hedge
against unhappiness after winning the lottery,
zombie-dinosaurs, sentient lightning, the flood.
There's a place on the border, in the demilitarized
zone, entire villages between troops tired of holding
weapons, counting on destruction to hold like a wall.
There's no way to save love in a bottle. The landscape
shimmers with each breath, each promise, each word,
the name for sky that will hold the weight, our faith.

Bodies Of Water

I grew up on Lake Huron,
wind scent inside everything
vespers to some aching shore.
The years brought the churn
of too-cold water to surface,
raised dunes from the waves.
We all read about vessels lost,
rough waves the size of living.
I navigated far, I lost myself,
my reinvention of dry riverbed
along a false trail of stone eggs,
other men's kingdoms, the salt
never able to rub away the lake.
I became a myth that I launched
every so often, from coral reefs
to the glaciers of the invaders
I carry in my blood and deep
recesses no one but me could
drown in. I was impossible.
It was impossible to foresee.
The lake flowed in your wake,
memories of slow submersion
were always there, in the fingers
of the tributaries we came to be.

Coming Of Age Poem Using Fifty Words That Might Cause The NSA To Flag You As A Terrorist

His mother would *sweep* her *mace* into an *indigo*
purse and *badger* him, "Slow-poke the *artichoke*,"
for preferring *Reno* to the college *snuffle, beef market*
of *lacrosse* tossers, Jell-O shots, and *credit card fraud.*
His sometimes *flame Jasmine* got him in the *zone*,
loin to *loin*, on the *basement* couch, their chosen *niche*,
utopia of *quiche* and *salsa*, his *red-headed Capricorn*,
quick to *unzip* for *sex*, and call his thrashing *fish* a minnow.
Did *keyhole eavesdropping* on his sister's friends bring *Ninja*
stealth? *Jack* told him to run, *Austin nerd*, full of *cocaine*
and malaise, afraid of *Texas*, and dropping *dead* from *blowfish*
darts, her *gorilla* boyfriends turned into *clandestine snipers.*
He suited up, *Roswell cowboy*, not afraid to strap on his big
asset, his *Macintosh*, to *face* the *fangs* of starving career *advisors*
peering at him like a *veggie* burger without French fries or bun,
the *enigma* of missing something as hard as missing none.

Bits

Sometimes an insect shears a limb and hops
to keep pace with the collective madness.
Sometime a child loses a parent or vice
versa and a doorknob clatters to the floor.
Sometimes a flower falls apart to keep
the bees confused and the wind in check.
Sometimes a planet escapes orbit
and cannot help itself from seeking the void.
Sometimes loss is a tunnel or a cave
and the place can change depending on the day.

The Imperfect War

The perfect poem exists,
a masterpiece from each era,
threading air, earth and us.
The perfect time exists
to strip the husk from seeds,
and swing in the boughs.
The perfect me exists,
but I've never met him
floating with hummingbirds.
The perfect peace exists
in the blackness of a sea
with no top or bottom.
The perfect argument exists
to explain moon eye, sun
eye, the mask of Minerva.
The perfect dance exists
between birth and finale,
bare feet on another shore.
The perfect fever exists
to destroy the contagion,
faces baked in shadow.
The perfect goodbye exists
only in the imagination.

Anonymous

The dead do not have their eyes
on the living. They remember glass
doors blocking their rise and borders
beneath beds and silkworms for hands.
The Gravel Diaries are passed stone
to bone, organs groaning into the past.
The dying do not divide their limbs
for the dead. No abyss or submarines,
no buttons pushed or soldiers clipped,
zippers for flesh, bitters straight up,
no communion for the conscripted,
the horizon an amputated kite string.
The living do not display callouses
for the dying. Catastrophe is not physical
the way a man forgets to crave ovens,
sour in heat and disdainful of lovers.
What do you wear for your embalming?
The box feels absolutely nothing inside,
saga of perishable containers carrying us.
The dead and the dying and the living
stop lying to the anonymous blossoms,
make amends to Orion, string olives
into rosaries and fling rice into eyes,
flee wings and mysteries, the vying
and the denying...oh so trying.

Riddle

A retired interrogator walks
into a bar with himself,
and asks for bold spirits,
untraceable in the lineage
of fevered fermentation.
Who is greater than gods,
creator of zealots and fools,
apocalypse of every shade,
architecture of storm and awe,
maker of mountainous tombs?
His drink is the hue of mist,
answers to each question
settling in the bottom
like the ghosts on stools
falling into tiny screens.
We don't know how to feel
until we read our own words,
each book a clue, a snakeskin,
to contemplate the uncharted,
cheer for the unctuous and unsung.
He thinks about a knock
knock jokes, but no one
wants to open that door,
identity a symbol of loss,
a joke that masks pain.
We knew the answer beforehand,
the question we all must address
in a time of makers and fakers,
the skeletons dipped in syllables,
in a garden of relative delights.
There was once a woman,
silhouette in departure,
perhaps lover or mother,
the trail into the unknown,
familiar voice in the distance.
The mission is to look inward,
to understand the clash and quiver,

and legendary mistakes on tongue,
for all that is left of this adventure
is what it means to believe in one.

Dangers Of The Road

Scientists tracked motorists by satellite
to see which of them would swerve
over the median to mow down small
animals, and many chose blood sport.
My friend Sarah told me how she
was terrorized by a stalker for years,
changing her address to flee this hole
of a man threatening to pull her in.
My sergeant once told me that killing
another is our passage into manhood
the same way a woman is wounded
giving birth, a screaming revelation.
Armadillos and lizards suffered equally,
with men in SUVs more likely to murder.
Sarah was walking on a secluded beach
when her terror popped up behind her.
I learned to point a rifle at an enemy
and to stick a bayonet in to the hilt.
Dangerous drivers approach us all.
She smashed in his skull with a rock.
Life began after my daughter was born.

The End Of The World Did Not Happen...

With children diving under desks
and clouds that looked like angry
balloon animals. From family members
eager to turn your outsides into their
insides. By a souped-up Commodore 128
computer that decided you could be
a delectable source of electricity.
Through a flood or emotions triggering
mutant powers. During an angry
volcano surprise party under Los Angeles,
angry gods, or angry paratroopers
descending on cornfields with machine
guns. Because of a failing sun or genius chimps
leapfrogging the evolutionary ladder.
Due to asteroids or ozone leaving the planet
with bald spots, tiny dictators in alligator
boots, or time-traveling paradoxes of crushed
monarchs and nations under altered beliefs.
But fear and ignorance could set the table,
a conversation or two, a desire to please,
ash covering the cars of coughing cities,
a step at a time, a castle at a time, a prayer
of blood sky and bowed heads. It could shift
with a moon-sized parable where the one
can save the many. It could be you this time.

Return To Mermaid

I returned to her from overseas,
unable to dunk away the desert
sands and unwashable stains.
She waited for me in the trailer
décor, a thrift-store museum
of mismatched boat relics, walls
lined with buoys, fish mounted
on the walls with mouths agape,
posters of famous ocean tragedies.
The carpeting was a polluted green
sea and the overcast stucco ceiling
a single cloud mass observing us.
My mermaid nested in the depth
and depression of me, a glimpse
of the man I drowned in fatigues.

You Can't Kiss An Astronaut On A Space Walk

You can't even get a proper
tan, solar winds be damned.
The pull is dark, irresistible,
a mission that could freeze
us beyond any border.
So much to discover from here.
You flew to me so many times,
and I'd stare at the picture
you sent me from Ipanema,
your sunglasses two moons,
helmet visor reflecting descent.
So much to miss in our orbits.

Unclaimed Baggage Center

In an airplane hanger
you rifle through the lives
of the newly missing,
massaging lost luggage,
mismatched hosiery,
tennis balls, designer
shoes, and track suits.
Where are the passengers
to the other side?
You fondle the scarf
with crocheted antlers,
team sweatshirt fraying
at the sleeves, electric
toothbrushes, bandanas,
and trinkets forged
in faraway factories.
How many are lost
in worlds of fabrication?
You fear to rummage
through battered purses
and briefcase because
of miniature Buddhas,
cross-eyed porcelain
dolls, half-masticated
gum, and twin castanets.
So many signs of loss
on the road to somewhere.
You have a vague
feeling if you search
hard enough you'll discover
Sasquatch, the missing link,
your childhood friends.
Hidden in the back
is a washing machine.
The stains of the missing
cannot be scoured away.
The lingering scent
is familiar, antiseptic,

like hospital corridors.
There is a miracle
department that houses
unused coffins, wheelchairs,
canes, and oxygen tanks.
How could we have taken
flight without them?
Unrequited objects remind
us of brittle homecoming,
desperate departures,
belief in the ways we
remember ourselves,
coils of unfurled wings,
the loneliness of hiding.

Core

The weight of the world is hidden
by context, no scale, no way to know.
The upset gut and gurgles underground
do not unearth the stress: kids, bills,
unsteady pitch. The screensaver is sky
and meadows. Fields waver in dying
light in a rhythm beyond time, past
brittle smiles. The weight of battle
cannot be measured in a busy day.
The heart is a feathery fossil. It used
to beat. It used to soar. Today is swollen
with need. Do you hear it? When tanks
quake, when hands shake, when hills
tumble at the desk. Courage is holding
the pieces and knowing there's more.

33 Lessons In Camouflage

1

Mom switched me from southpaw
to righty, an abandonment of sorts,
child on a basket of reeds adopted
by a land too scoured to hide beliefs.

2

As a boy, I read a fable of stacked
turtles stretching to touch the sky,
moral lost, just the teetering way
of latching onto the ladder upward.

3

A pigeon on the Hollywood Walk
of Fame picks at a discarded fried
chicken bag, cannibalism a lesson
of survival when scuttling on stars.

4

My ghost hand reasserts itself in odd
ways, picking weeds, steering wheel,
pleasing a lover, terrapin unshelled
when there is no one to drop you.

5

Interrogation was a nesting doll
of camouflage, uniform as skin,
questions as fists to pummel any
chance to unhinge the man within.

6

The checkered shirts in my office,
flimsy remnants of boyhood flannel,
hide the wolves and sheep in equal
measure, demand wrapped in leisure.

7

The coyote outside my window curls
into the dirt, family scattered beneath
sloping canyon, the hunting reunions
enough to wake me even in daylight.

8

The melancholy house with an address
ending in ½ displays a blue door, one per
floor, opening to air, cerulean patches,
no clouds, inner tranquility or madness.

9

I lost my glasses once in a snow pile,
too scared to tell my folks, sifting
powder day by day until thaw, glad
for once they didn't notice my loss.

10.

Good interrogators feel more not less;
it's not nearly enough to wear a mask,
eye mirrors projecting wish fulfillment,
ricochets inflicting collateral damage.

11

I once read a poem about a butchered
goat and told myself that it was about
the war, the ritualistic flaps and knife
sharpening, the earth soaked through.

12

Our gladiators have cameras on them.
They strike in unexpected flurries.
The stadiums vibrate with victory
chants. The ballet of the broken.

13

A couple in my neighborhood hangs
chandeliers in trees, beacons to kids,
the illumination of the outside path,
sorrows pushed onto the lawn.

14

One day, in a hurry, I felt an object
in my shoe and discovered a button
from my shirt, remnants of a love
affair of objects behind my back.

15

We scrubbed the tracks of armored
vehicles for hours after mud runs,
separation of weapons and earth,
the union of men in boiling hours.

16

The world's most expensive coffee
is sifted from poo, Jacu Bird brew,
Black Ivory blend, civets force fed
plump beans, a filter of pit and rue.

17

My boss used to ask me to smile more,
a daily dose of subservience, white
gates opening, the lesson of keeping
enemies inside without swallowing.

18

In dead planets drifting in space,
between solar freeze and magma
core, life could coil
in countless oceans in the dark.

19

When my girlfriend takes a shower
I stare at her silhouette through curtains,
the outer sheath a filmy map of the world,
breath melting ice caps, my own thaw unseen.

20

The first person to nestle a baby
in a cake, in any form, must have
known bulges cannot be hidden,
that hunger is its own discovery.

21

We started a fire on the shooting range,
beat the brush with ponchos to keep
the forest from engulfing us, tiny enemies
licking at our feet, the wildfires of war.

22

The state of being is not unlike
the state of unbeing. The illnesses
that writhe beneath the surface pull on us,
tiny passengers with unknown agendas.

23

The 25th hour is like an earring
worn in an experimental phase,
the past spilling from corduroy pockets,
once baggage in the uniformed me.

24

The older I get, the less well I do at hide
and seek, my kids able to see the bulges
poking out, fewer places for me to disappear,
the essence of fatherhood to be in plain view.

25

The pop star jokes how a girl who hid during
the war would have been a screaming
fan, his vanguard planning an invasion,
the truth of the man strewn in the message.

26

My daughter and I notice the missing letters
on the pastry signage: *Do_uts* and the other
homonymous messages of warnings, the truth
of LA visible not in beauty but the fading away.

27

There is a parasite in cats borne from rats that drives
risky behavior. Nearly half the world's people
have the urge, so many of our compulsions wired
and wrung, the path to our nature kinked and knotted.

28

When I was a boy I read the old and new testaments
on my own and knew that God was a jealous
bully, the bible the word of men I did not want
to follow, faith in myself: divinity in strange places.

29

The past harmonizes, waiting for when
we are most like our former selves, a symphony
that thrums in foot tapping and humming,
the body a radio for ghosts we've forgotten.

30

I played army with the box of men kept
alive in the attic, soldiers in battles
with cowboys and dinosaurs,
an alien armada waiting, a man waiting.

31

We lose ourselves in books, in love making,
in the crannies of our work and passions,
the miscellany of the unrecorded world,
discovery of who we choose to master.

32

Not long ago we placed men in boxes, iced
them, drowned them, for the idea of freedom.
The measurement of pain is international.
Not long ago is idling down many roads.

33

I hide my strengths and weaknesses,
clever boy, but my children expose
them with their own. Each day a scab
is torn and each night a new me forms.

Acknowledgments

The Alembic / "Unclaimed Baggage Center"
The Antioch Review / "Prayer for Morning Commute"
Asimov's Science Fiction / "Stranger"
Bearers of Distance: Poems by Runners / "Mile Post"
Bluestem Magazine / "Shelter"
Booth / "Dangers of the Road" and "Why I Worry My Mother is Dying,
    Explained by Five Extinct Punctuation Marks"
The Café Review / "The Gravel Diaries"
The Cossack Review / "Morels"
Devouring the Green: Fear of a Human Planet: Poetry Anthology / "I Lost
    the Robot in the Divorce"
Dear Cancer...The Anthology / "Morels"
diode / "Calling the Time", "Policy", and "To the Guy Who Drew a
    Penis in the Elevator"
Epoch / "Why My Father Carries Three Guns"
The Evansville Review / "Marks"
failbetter.com / "You Can't Kiss an Astronaut on a Space Walk"
Jam Tarts / "The Island of Escalators"
Kentucky Review / "The Human Library"
The Los Angeles Review / "33 Lessons in Camouflage"
The McNeese Review / "Imagining the Mouth of My Lover Holds the
    Power of Good and Evil"
New Verse News / "Pilates Instructor Calls 911 after Mistaking Labra-
    doodle for a Lion" and "Coming of Age Poem Using Fifty
    Words that Might Cause the NSA to Flag You as a Terrorist"
Notre Dame Review / "The King of Camouflage" and "Invisible Monsters"
Prelude Magazine / "Mr. Old Year"
Prairie Gold: an Anthology of the American Heartland / "Bodies of Water"
Rise Up Review / "Why My Father Carries Three Guns"
Tampa Review / "The Imperfect War"
Then and if / "The Mystery of Three Things"
Till the Tide: An Anthology of Mermaid Poetry / "Return to Mermaid"
Wherewithal / "Anonymous"

About the Author

A former US Army interrogator and longtime resident of Los Angeles, Martin Ott has published eight books of poetry and fiction, most recently *Lessons in Camouflage* (C&R Press, 2018). His first two poetry collections won the De Novo and Sandeen Prizes. His work has appeared in more than two hundred magazines and fifteen anthologies. Content is Martin's passion – he works as a marketing communications professional and develops for TV and film in between other writing projects.

OTHER C&R PRESS TITLES

NONFICTION

Women in the Literary Landscape by Doris Weatherford, et al

FICTION

Made by Mary by Laura Catherine Brown
Ivy vs. Dogg by Brian Leung
While You Were Gone by Sybil Baker
Cloud Diary by Steve Mitchell
Spectrum by Martin Ott
That Man in Our Lives by Xu Xi

SHORT FICTION

Notes From the Mother Tongue by An Tran
The Protester Has Been Released by Janet Sarbanes

ESSAY AND CREATIVE NONFICTION

Immigration Essays by Sybil Baker
Je suis l'autre: Essays and Interrogations by Kristina Marie Darling
Death of Art by Chris Campanioni

POETRY

Dark Horse by Kristina Marie Darling
All My Heroes are Broke by Ariel Francisco
Holdfast by Christian Anton Gerard
Ex Domestica by E.G. Cunningham
Like Lesser Gods by Bruce McEver
Notes from the Negro Side of the Moon by Earl Braggs
Imagine Not Drowning by Kelli Allen
Notes to the Beloved by Michelle Bitting
Free Boat: Collected Lies and Love Poems by John Reed
Les Fauves by Barbara Crooker
Tall as You are Tall Between Them by Annie Christain
The Couple Who Fell to Earth by Michelle Bitting

CHAPBOOKS

Atypical Cells of Undetermined Significance by Brenna Womer
On Innacuracy by Joe Manning
Heredity and Other Inventions by Sharona Muir
Love Undefind by Jonathan Katz
Cunstruck by Kate Northrop
Ugly Love (Notes from the Negro Side Moon) by Earl Braggs
A Hunger Called Music: A Verse History in Black Music
by Meredith Nnoka

www.ingramcontent.com/pod-product-compliance
Lightning Source LLC
Chambersburg PA
CBHW031153090426

42738CB00008B/1307